THE JAMES BACKHOUSE LECTURES

I0026128

This is one of a series of annual lectures which began in 1964 when Australia Yearly Meeting of the Religious Society of Friends was first established.

The lecture is named after James Backhouse, who travelled with his companion George Washington Walker throughout the Australian colonies from 1832 to 1838.

Backhouse and Walker were English Quakers who came to Australia with a particular concern for social justice. Having connections to social reform movements in the early colonies as well as in Britain, Backhouse and Walker planned to record their observations and make recommendations for positive change where needed.

Detailed observations were made of all the prisons and institutions visited by Backhouse and Walker. Their reports, submitted to local as well as British authorities, made recommendations for legislative reform. Many of the changes they initiated resulted in improvements to the health and wellbeing of convicts, Aboriginal people and the general population.

A naturalist and a botanist, James Backhouse is remembered also for his detailed accounts of native vegetation which were later published.

James Backhouse was welcomed by isolated communities and Friends throughout the colonies. He shared with all his concern for social justice and encouraged others in their faith. A number of Quaker meetings began as a result of his visit.

Australian Friends hope that these lectures, which reflect the experiences and ongoing concerns of Friends, may offer fresh insight and be a source of inspiration.

With a continuing COVID19 and winter flu epidemic, the 2022 Backhouse lecture Working for Justice in a Warming World by Yarrow Goodley, a Member of South Australia and Northern Territory Regional Meeting, was presented online via Zoom to a wide audience on July 3rd 2022.

Ann Zubrick
Presiding Clerk
July 2022

Quakers
AUSTRALIA

THE JAMES BACKHOUSE LECTURES

2022

THE **JAMES BACKHOUSE** LECTURE

Creating hope: Working for justice in catastrophic times

Yarrow Goodley

Quakers
AUSTRALIA

© 2022 Religious Society of Friends (Quakers) in Australia
PO Box 4035 Carlingford North NSW 2118
secretary@quakersaustralia.info
quakersaustralia.org.au

ISBN 978-1922830-09-8 (PB); ISBN 978-1922830-10-4 (eBk)

Design & layout by:
Interactive Publications, Carindale, Queensland, Australia

Cover image: *Upcycled shirt, mended, embroidered and hand-dyed* by Mary Heath, one of the 2022 Backhouse conversations cohort.

Contents

About the author

Yarrow Goodley was nineteen years old in 1988 when the Intergovernmental Panel on Climate Change was founded, and the IPCC's five-yearly reports have sounded ever more dire warnings throughout their adulthood. Yarrow began working with young children in 1996, and has been an activist educator ever since, working with babies, toddlers, preschoolers as well as adults studying or working in early childhood education and care. Yarrow's doctoral work was in the sociology of education, and their book *The Sociology of Early Childhood: Young Children's Lives and Worlds* (2019) explores how inequality continues to shape and reshape the lives of young children. As an early childhood educator, Yarrow is reminded every day of the uncertain future that awaits our youngest citizens. These children will be Yarrow's current age in 2070—a future that may be either apocalyptic or utopian, depending on our actions now. As a Quaker, an activist, and a gardener, Yarrow aims for that utopian future, even when the path to that place is murky.

Introduction

When I was asked by the Backhouse Lecture Committee whether I would do this, I felt immediately dismayed. I wanted to look over my shoulder to catch a glimpse of the person they were really talking to. Surely, this was a job for a spiritual person, a wise person, or an elder. I wanted to say 'no', and there have often been days when I wished I had. A clearness committee helped me feel my way through these conflicted thoughts, to perceive more plainly how this response came out of years of self-doubt and self-hatred. Although I often feel like a broken person, my Quaker faith reminds me that I have an inner light, as do each and every one of you, and that if I seek this inner wisdom and clarity, then I can indeed speak to my experience of the truth. As I tell this story, I will include moments of silence, honouring our Quaker practice of using silence to wait for and observe our own inner light, or voice. Do not be alarmed by these silences—but rest in them, knowing that more words will be on their way.

There have been many who journeyed with me along the way, including writers, activists, friends and loved ones, and the members of that clearness committee who helped this to happen. I hope you can join me on the final steps of this journey, as I share with you my reflections on the climate crisis. We can and must use this key moment in history, as the prophet Micah's words are reported: '[God] has told you, O mortal, what is good; and what does the LORD require of you but to do justice, and to love kindness, and to walk humbly with your God?'. How can justice, kindness and humility tell a new story about the future of our beautiful planet?

Whose story am I telling?

In one very real sense, the only story I can tell in this lecture is my own. I am Yarrow, a Quaker, an early childhood teacher, a sociologist, a non-binary and queer activist, and a gardener. I could give myself many more labels, but these are some of those I feel most deeply and also those most relevant to what I am saying today.

I am a person of considerable privilege. I have white skin, which continues to carry an incredible amount of privilege, wherever I am in the world. I will be less likely to be poor, to be imprisoned, to be rejected as an asylum seeker, or to be insecurely housed. I will be free to pretend racism, either personal or structural racism, does not exist, and in fact most of these privileges will be invisible to me.

Most of the people I need to interact with, especially those in power, will have white skin, and will trust me more, and treat me better for no other reason than my whiteness.

I was assigned the label 'male' at birth, and, although this label has rarely made any sense to me, it has still afforded me a great deal of privilege. I have been less likely to be sexually assaulted or harassed growing up. I have presumably been more likely to be preferred when applying for work, and will earn more when I do. My views will be treated with more respect, and my voice listened to more often, regardless of the clarity or wisdom of my opinions.

I was born in the UK and brought to Australia as a child, meaning I am a citizen of two wealthy nations, and have been granted many rights, some formal, some not, as, a result. I have the right to vote, to low-cost healthcare, to affordable education, and to freedom of thought and expression because of where I have been born and lived. The wealth of these countries means I have lived a luxurious life compared to most of the people on this globe, with access to clean water, abundant food, a safe and secure society, and a relatively clean environment.

I was raised middle-class, meaning I have been protected from most of the harshest impacts of poverty for most of my life. I have been taught to value education, and been supported to access it at all levels, and with confidence, and know that those around me will view this positively, rather than with mistrust.

These things I did not always know, and so many of my thought patterns and behaviours growing up were shaped by these unearned forms of privilege. It was only as an adult, and through my training as an academic sociologist, as well as my involvement in feminist, queer, and climate activism, that gave me some insight into this privilege, and continue to help me in unlearning some of the least helpful of these patterns.

I feel incredibly fortunate in the present day, to be able to identify as non-binary, and have that be understandable to at least some of those around me. I can have this status recognised legally by state and federal governments, and even receive healthcare that affirms this identity, if I look hard enough. I did not have this experience growing up, because this identity would not exist in any meaningful way for the first forty years of my life. I felt confused and wrong for most of my life about how I should act and feel, and continue to flinch internally whenever others make assumptions about my gender, and thus who I am likely to be. Nonetheless I feel fortunate that I have had this experience because it has helped me understand in some ways the *absence* of privilege, and so have some empathy towards the majority of my fellow human-beings, for whom gender, race, Indigeneity, class, disability or age may be held against them.

Can this story be trusted?

Chinua Achebe, the Nigerian novelist, brought to the world's attention an old

African proverb: 'Until the lions have their own historians, the history of the hunt will always glorify the hunter'. This is a useful reminder of the distortions in understanding that occur as a result of unearned privilege. How can I hope to see this climate crisis clearly, when my vantage point hides from me most of the problems that it causes? As someone who has a substantial amount of privilege, the importance of climate *justice* has been hidden from me for most of my time as a climate activist. Although I probably started becoming an environmental activist as I came into adulthood in the late eighties, it was not until perhaps the last five years that climate justice has come into focus for me. In order to tell a more complete and nuanced story, I will be drawing throughout this lecture on a series of conversations I have had with climate activists from within and beyond our Quaker community. These activists were drawn from my own networks, and so represent only a thin slice of human diversity. However, I tried to choose people whose experience of the climate crisis would be different from my own, due to their age, their expertise, or their life experiences.

I will allow them to introduce themselves now:

> *My name is Rowe. I was born with a profound love of the earth and all nature, which led to running away or refusing to be brought inside for much of my childhood. This love has endured, strengthened and deepened. When I met cosmology, I was enamoured and loved it as my religion of miracles although it is science. Permaculture framed my life's activities around the Earth and its restoration, while Quaker beliefs are still my foundation. They are deeper now under different stresses and times. They are inexhaustible. I love reading, enjoy film, want to walk big distances and garden more before frail age stops me. I want to spend time with people I value.*

> Rowe – "*Understanding you were given life against all odds.*"

> *My name is Lynn Lobo. I prefer the pronouns they/she. My cultural roots are in Goa, India. I am an awareness facilitator, anti-racist trainer and climate change activist. For me, climate change and racism are inextricably entwined within the bedrock of our capitalist system. I've had a long career as an Acupuncturist and Psychotherapist. These days I teach how to work with dreams and body symptoms within an M.A. program on conflict facilitation at the Process Work Institute, Portland, Oregon, USA. A Daoist at heart, I love meditating, painting and losing myself in nature.*

> Lynn – "*The earth is dreaming through us,
> and collectively we are dreaming together.*"

> *My name is Garth. I've had a disabling chronic illness (ME/CFS) near-*

3

ly as long as I've been striving for climate justice and against growthism (about fifteen years). An erstwhile software engineer, now agronomist, my activism has been focussed on developing local community, cooperative enterprises, demonstrating alternative lifestyles and values, and exploring the possibilities of new economic and social systems. Nowadays I am mostly housebound, so my contribution is limited to emotional support and cheerleading for my extraordinary wife.

Garth – *"These are not individual choices; these are systemic issues."*

My name is Elizabeth. I see myself as an emerging responder (in emergencies and quieter roles) to the unfolding crises of this age of human-formed devastation and complexity, that also has much joy throughout. I value finding ways wherever I can, through wild gardening, art and poetry, healing and bodywork, journalism and radio work, community support for events and activism, meditation, friendship, and alchemy of the kitchen. My concerns of climate change and earth laws including the abolition of nuclear weapons (and all destructive tech) exercise me. Beauty, light and shadow, suffering and wholeness are my informants at present. Emergent, indeed.

Elizabeth – *"Where we are now encompasses all of time."*

My name is Monica and I met Yarrow through Rise Up Singing, a community singing group, learning, rehearsing and singing songs about climate change for Climate Actions—street marches, rallies and events. I started being an activist as a teenager when I stood up in church to object to the Priest saying that God wanted us to fight against communists in Vietnam. Since then, I have been active as a feminist, a lesbian, conservationist, a social justice and climate justice campaigner. From the Franklin River to Extinction Rebellion, I have continued to demand change. For the past twenty years, I have been supporting refugees who have been treated so badly by our governments.

Monica – *"It's feeling very deeply connected to the Earth."*

My name is Millie, and I am the National Coordinator for Australia re-MADE. I have a qualitative research background and have spoken in-depth with hundreds of Australians about their lives, communities and dreams. I've worked in and around universities for over a decade, building student capacity and enthusiasm for tackling wicked problems. I am also a carer for my family and community and am passionate about acknowledging this work as a valid, valuable and legitimate use of my time. I dream of becoming a butterfly dancer on stilts.

4

Millie – "*Networks and community are the infrastructure of the revolution*"

My name is Mary. I am a climate activist working with Extinction Rebellion. I am a former professor of law. I left that job in order to be able to commit more time to climate activism. I also teach mending, guerrilla garden—planting native species where only weeds and rubbish are growing, and enjoy my own garden. I began my activist life in the peace movement and have had a lifelong commitment to feminist and other forms of activism, including campaigning for a rape-free world and participation in the trade union movement and the campaign opposing a bridge to Kumarangk (Hindmarsh Island).

Mary – "*A community response is the only one that makes any sense to me.*"

My name is Gerry. I know myself as a being who is grown out of and held within the earth. I am led to act out of this knowing. I am a partner, father, grandfather, in a wonderfully blended family. Via Thomas Berry and cosmology, I found my home in earth spirituality. This has grown through the ongoing practice of holding space for difference through shared silent worship with Friends and immersion in the writings of such as Joanna Macy, Arne Ness, many science writers, and the later writing of John Yungblut. I live on Dja Dja Wurrung country—the rocks are my grounding and the mosses, lichens and fungi that abound here are my Elders.

Gerry – "*If we don't change those [our political and economic systems] we're going to continue hitting the wall again and again.*"

My name is Peri. My earliest memory is standing in a saltmarsh in France, watching strings of geese flying south while the cordgrass bowed their heads at their passing. That set my feet on the path—I studied botany and environmental subjects to learn more about the wettish world of swamps, fens, bogs and marshes. The bowl of the sky in such places set a young mind to enquiring on infinity and eternity. I am not primarily an activist, but when people can't hear the voice of the world, someone must speak up. I have worked with Elders protecting Fraser Island, students protecting forests in SW Australia and Tasmania, and defending refugees. But now the threat is existential, so I accompany my grandchildren as they speak up for a suffering planet.

Peri - "*These ecosystems don't read the Sunday papers – they just pick up their skirts and move.*"

My name is Habibah and my climate activism is deeply rooted in my practice of Islam and my fight for making the world a more equal and just place. I see the exploiting, plundering and abusing of the planet, of nature, of natural resources and of the environment as a system that applies these measures on humans as well. I am a revolutionary socialist, and I believe that the entire system must change for us to achieve climate justice. I seek to use the path of journalism to facilitate conversations and to organise movements that are my cause. I love to connect with the ocean the most because the vastness and the constant movement of the ocean reminds me of the cycle of life and how precious it is.

Habibah – *"Climate justice is racial justice."*

The threads of all of their stories weave through my own, creating a stronger fabric than it would be if it were just my own experiences and perceptions.

– silence –

Climate injustice

Climate justice has been explained insightfully by Hop Hopkins, from the Sierra Club (one of the world's oldest environmental organisations, founded in 1892). As Hopkins (2020) explains, in an article entitled "Racism is Killing the Planet", 'you can't have climate change without sacrifice zones, and you can't have sacrifice zones without disposable people, and you can't have disposable people without racism.' When multinational fossil fuel companies and other extractive industries are recklessly and wilfully changing our planet's climate, they are doing so by illegitimately possessing First Nations' land, and locating their polluting industries in places where poor, black and brown people live, and the consequences of that global heating is felt first and hardest by the world's majority, in those countries which have been made least resilient by the long-lasting damage of colonialism.

Later in this narrative, I will tell you stories about climate justice, but for now I will try and paint a picture of climate injustice, and the ways in which it perpetuates itself. To talk about climate *injustice* is to open our eyes to the horrifying realisation that not only are we damaging the precious ecosystems of this planet, but in the process we are exacerbating almost every dimension of injustice on this planet.

As one of my climate conversationalists, Mary, explains it,

> I think of it as an approach to understanding and responding to the climate crisis that puts the voices, perspectives and experiences of people who are at the front line, and most vulnerable, at the heart of how we understand the climate crisis and how we choose to respond to it. For example, that means if we're not thinking about how this crisis will affect... Indigenous and First Nations peoples, who've been on the front line of this crisis all along, and have been at the front line of trying to protect the natural world... forever, and certainly since colonisation and capitalism came along, which was a huge crisis for all First Nations, everywhere, then our perspective is not much good. If we're not putting the perspectives of young people who are disproportionately going to live with the outcomes of this, by comparison with myself in my mid-fifties, then there's a problem. If we're not thinking about people who are already poor, the global poor, as well as the local poor, if we're not thinking about them first and foremost, then we're not addressing the problem. And actually, if we're not thinking about people whose jobs depend on extractive industries, then we're crippling our capacity to move forward.

Mary's insights resonate with Hopkins' view, in thinking about the people who are currently disposable in a world ignoring this crisis—the poor, the young, First Nations—who all see their vulnerability clearly and are trying to get the world's attention.

Historical origins of climate injustice

It is likely that we have been changing the climate for a long, long time. Some scientists see the beginnings of this in humankind's shift into agriculture and domestication of animals. This seismic shift in human culture led to the cutting down of forests, which once covered nearly all the continental land masses below the snowline. This process began almost imperceptibly, with human populations perhaps numbering only a few million across the world. Yet because productive agricultural land was the main source of wealth and power, the incentives to cut down trees were strong then, and even now deforestation is one of the prime drivers of our changing climate, including here in Australia. It may be that our Holocene age, this period of mild climate between ice-ages, has been extended unnaturally by our climate changing, with a global ice age being long overdue in geological terms.

For many thousands of years, deforestation was humanity's only contribution to climate change. Many human communities, including those we now see as First Nations communities, developed systems and cultures that aimed to preserve and protect the land, which was and is viewed as home, mother, and spiritual foundation. Yet some human cultures, including those at the root of what are now thought of as 'Western' nations, developed spiritual systems that challenged the notion that the land is sacred. From the days of the Greek and Roman empires, the Egyptian empire and perhaps the Mayan empire, some human communities have reached out to conquer and subjugate neighbouring countries, creating legendary stories of power and wealth, and hidden stories of damaged climates and eco-systems. Even before the massive damage caused by modern weapons, early human wars saw crops burnt, farmland salted so as to become useless, and forests pushed back to create defensible zones and military routes. We know it is likely that these early waves of colonisation also spread plants and domesticated animals into new places, and some of these would have become endemic, disrupting and changing local ecosystems. Legendary stories of these powerful conquerors—Alexander the Great, Julius Caesar, Ramesses the Great—created envy amongst later rulers, driving a desire to emulate their achievements ... and their destructiveness.

As we know only too well, conquest and warfare create conquered populations, and slavery was a widespread practice across those cultures. As well as enslavement, conquered populations create resentment and revolt, and become the cause of future wars and rebellions. As Quakers, we have a long-standing opposition to both war and slavery, knowing that both of these treat some people as dispensable, rather than as a part of the divine, a reflection of God. The story of climate injustice I am telling today understands that those catastrophic and hateful practices are part and parcel of a larger story of injustice that has been damaging not just human beings, but also other-than-human creatures, plants and fungal communities.

The modern era

Fast-forwarding a couple of thousand years, we find our human capacity for innovation embracing new thinking, and different forms of knowledge. This was the time of the reformation, revolutions, and the Renaissance—all times of change, or re-examining our place in the world. Most religions are inherently conservative (including Quakerism, although we try to believe otherwise) and while religion and the divine right of kings prevailed, societal change was slow, disrupted only by war and conquests on a local scale. In challenging the power of religion, rulers were overthrown, and new systems of thought were created, accompanied by a massive expansion in science and technology. Advances in ship-building enabled the crossing of oceans once thought to be endless, and technologies such as gunpowder tipped the scales heavily in favour of those who could acquire them. A new era of colonialism was born, whose ramifications continue to haunt all of us, both colonised and colonisers. These conquests, mostly by people with pale skin over people with black or brown skin, led to a massive expansion of slavery, and the transport of millions of African people to the Americas and beyond. While forms of racism have probably been with us forever, this was a time when science sought to justify these crimes against humanity through 'proving' that those being conquered and enslaved were barely human at all. Systemic racism was born, and continues to be maintained in multiple ways to benefit people who look like me, or who can pass as white.

The expansion of scientific thinking led to a rapid expansion of new technologies, and an industrial revolution. This involved a social revolution, with many leaving rural areas to work in factories in the city. Those early days of industrialisation had few or no regulations. Workers were killed in machines they were operating. Pollution flowed freely from factories into the streets and so into local watercourses, lakes and seas. This was the era of coal, which was burnt in steam engines, in domestic fireplaces, and to power many of these new factories. Mining these natural resources—not just coal, but also metals, rock and sand—disrupted and polluted ecosystems and drinking water. It led directly to the deaths of many miners, and indirectly of the communities in which they lived, from diseases like black lung. Then, as now, those living nearest to these sources of pollution were the poor, with this poverty causing sickness and stunting development of their children. The conditions of crowded slums also exacerbated early epidemics, such as cholera and smallpox, leading to further misery.

This was the era when capitalism supplanted feudalism, with factory owners wielding immense power over their workers, and extracting value from their labour without any concept of fair wages. Extractive capitalism has been tightening its grip ever since, as our society's dependence on both fossil fuels, and the machines that they drive, continues to entrench their power. Having accumulated vast profits, without being held responsible for any of the externalised costs, such as environmental damage, these fossil fuel giants now spend significant sums distorting democracy to maintain their destructive practices.

This era of revolutions and industrialisation was not all doom and gloom. As a human culture we have struggled towards better forms of governance, with tools such as democracy, systems of legal precedent and independent judiciaries, and the ongoing extension of human rights. These tools have allowed activists to wrestle with many aspects of injustice, including climate injustice, as seen in the legal cases regarding climate pollution being taken up in modern times. One current campaign, which has gained momentum in the last five years, is to make ecocide—the wilful destruction of ecosystems—a crime punishable by the International Criminal Court. This campaign also works with countries globally to make ecocide a crime at national and regional levels.

Our governments now have considerable power—people power—if only they were willing to wield it. We saw a graphic demonstration of this during the early days of the pandemic, when the worst forms of poverty and homelessness were virtually eradicated at the stroke of a pen, with a doubling of social security payments, and the provision of hotel accommodation to those sleeping rough. There is much our governments can do, if they would only choose to take action on behalf of the powerless.

Yet there have been huge and consequential mistakes made in the name of 'progress'. For the first half of the twentieth century, most urban development followed public transportation, an urban form that tends to reinforce equality, as well as minimising the carbon footprint of travel. But, in the middle of the last century, policy-makers and governments in most countries allowed themselves to be beguiled by car manufacturers and freeway-building companies into skewing urban development towards cars, actively shutting down or underfunding public transport networks, such as the tram networks that existed in many of our state capitals. This has had multiple ill-effects. It encourages urban sprawl, destroying valuable agricultural or wild lands, as developers buy land on the urban fringe, and then bribe governments to rezone it, and then service it at taxpayer expense, in the interest of their profits. It has seen the rise of the private car, a major cause of climate pollution, as well as a driver of vast inefficiencies in land use. Freeways take about 20 times as much space to transport the same amount of people as a heavy rail line, as well as endangering the lives of those living nearby with the pollution from exhaust fumes. Car-based development means that around 15-20% of land in cities becomes devoted to car infrastructure, whether that be roads, parking, or buildings such as service stations. Urban sprawl actively disadvantages the poor, forcing those on the cheaper outskirts of cities to spend more on private transport, while living with substandard services.

Another vast mistake remains invisible to the naked eye, in the laws made to establish corporations across most wealthy countries. These have hothoused the growth of extractive capitalism, creating fictions of legal personhood which encourage businesses to pursue profit at all costs, with no legal consequences being borne by their owners or decision-makers, no matter how wrong-headed or destructive their decisions. These legal rules, and the systems of economic thought that were developed to defend them, have led directly to the immense power of multi-national companies, whose power and wealth often rivals or exceeds that of the countries they operate in. This becomes a modern form of colonialism, where the damage done by corporations is felt by the powerless—the people of the Niger Delta, First Nations communities in the outback, the textile workers of Bangladesh, to name just a few—while those who reap the benefits get to live in places of social, political and economic stability, pursuing lifestyles which produce carbon pollution equivalent to that produced by hundreds or thousands of less wealthy people. These economic theories and legal protectionisms ignore the glaring reality that our beautiful planet is finite, and that growth, and its accompanying pollution, cannot expand indefinitely (although some seem to see the moon or Mars as the next big real estate opportunity!)

Some climate injustice is deliberate, in the exploitation of Majority world workers, or the off-shoring of pollution production. Other effects are more accidental, but just as consequential. Changing climates destabilise countries, leading to war and conflict, exacerbating the flow of refugees, as happened most clearly in Syria and Afghanistan. The flow of refugees then has the potential to increase racism and protectionism in those countries that they are fleeing to, as has happened in the last couple of decades in Australia.

The overheating of our planet creates a need for better housing, to mitigate climate extremes, but it is the poorest people who can least afford to access this housing, many of whom are poor due to age, racism, dispossession, disability, or other forms of discrimination. Indeed, there are many people who lost their homes in the 2019-2020 bushfires who were living regionally because this was where they could afford to live, and who are still struggling to pull their lives back together. These climate impacts are real, and they are happening now, whether close to home or across the globe. Millie told me that the climate crisis became starkly real for her, when being forced to flee from those bushfires. As she recalls,

For the last ten years I've been anxious every summer ... but the summer, the 2019-2020 summer when we got evacuated from Tathra, was like ... I think for many people, like ... we were literally breathing in the ashes of ... dying things, and I think that was the experience ... We were always safe, we were never any of the people who were really stuck, but we evacuated that night, on New Year's Eve night, in this apocalyptic thick smoke and then had to make a dash to Canberra on New Year's Day before the highway closed.

Millie is keenly aware that scenarios like that will be repeated again and again, whatever our efforts to alleviate the crisis, as they have in places like California since that summer.

Our entanglement in climate injustices

As Quakers in Australia, we are a predominantly—though not exclusively—white and middle-class group of people. We do not want to think about our privilege, and so we frequently focus instead on our testimonies to equality, simplicity and peace, or our long-time commitment to many justice movements. Some of this privilege is a quirk of history, with early Quakers being excluded from the professions and other lines of work, and so being forced to become businesspeople. This also came about through our reputation for honesty and sobriety, but the end result has been an increasing level of affluence among our membership, and the comfort and complacency that accompanies that, even as we fail to notice it. Similarly, though we feel pride in our early activism against the slave trade, we did not in the end draw many ex-slaves into our Meetings and our friendship circles. Modern Quakerism has become more and more head-based, with high levels of education becoming the norm across our membership. This is not a terrible failing, but it can end up acting as a stumbling block to those who may have faced barriers to education throughout their own lives, such as First Nations people in Australia.

In a similar way, the climate movement has become dominated by white people, mostly because it is those who look like me who have the freedom from economic hardship, the education, and the available time to participate in climate activism. We are also the people who mainstream media pay most attention to, further entrenching this problem. This is not because people of colour do not care about climate activism. As Mary Annaïse Heglar observes, 'it's not just time to talk about climate—it's time to talk about it as the Black issue it is' (2020). As she explains, 'Climate change takes any problem you already had, any threat you were already under, and multiples it.' Problems such as systemic racism, disproportionate incarceration, state violence (such as with the *Stolen Generations*) and the financial struggles that often result from other inequalities are all being made worse by this ongoing crisis.

– silence –

12

Despair

Most climate activists become familiar with despair, and I am no exception. But this is not just my own story. Each of the conversations I had with other activists was an enriching experience, helping me sit with the issues I was exploring, and understand more deeply how many ways there are to step into the activist life. Within this extended community I could understand better how injustice perpetuates itself in a broken world. This injustice has many impacts, and those I spoke to during my preparation for this talk, described their own journeys into and out of despair.

Spiritual impacts

One of the impacts despair has is on our spirit, and our sense of spiritual connection. Much of this is based on a deep understanding of the Earth as sacred. To see it continually being degraded, polluted, or overharvested can feel like our world suffers from wilful blindness to the awe-inspiring complexity of our planetary home. A number of those I spoke with understood their despair in spiritual terms, and drew on their faith to uphold them in those moments.

Peri says,

> I'd be lying if I said that I didn't have moments of despair. And I don't always deal with it as well as I should. You know, sometimes I just get ... unbelievably angry ... There are days when, when I see the oblivion of people to what's happening, and it's infuriating. But mostly ... what I do is go out in the saltmarshes and sit and have a listen. You know, you can reach that of God anywhere, you just have to stop ... It really doesn't matter where you are. And if you stop and listen you will hear the little stirrings of the world around you. And you know, the world is working so hard to pull itself back into equilibrium.

Habibah has come to a similar conclusion when despair strikes. As she explains,

> I do have moments of despair ... and I have been learning to sit with it, instead of immediately trying to distract myself from it ... because it comes with a lesson—it doesn't come on its own, and it's not shallow—I've had to learn a lesson, and learn how to sit with it.

Despair always challenges us spiritually because it seems to reveal the hollowness of what we believe and how we are living. However, we can take away its power by sitting with that feeling, acknowledging it, and seeking to find the truths beyond that, the lessons it may offer, such as the incredible regenerative capacity of the earth which Peri sees so clearly.

Elizabeth was sceptical about both hope and despair, seeing both of them as something of an illusion. On the subject of despair she reflected,

> *There is behind despair a rightness about the awfulness that we are witnessing, because of everything that has gone before, that we, as humans, have created. Hope and despair are not separate from each other, in that regard ... they both represent a wholeness ... I know that there's a huge redress that is required to make the Earth anywhere near habitable for the foreseeable future.*

In seeing through the false comfort of hope and the apathy of despair, Elizabeth instead chooses to notice what is needed—a healing of planetary ecosystems.

Emotional impacts

For some people, their despair appears to hit them at an emotional level. Monica notices that what are sold as 'climate solutions' are often far from the answer. She realises, 'I'm gonna die, and I might as well go sooner rather than later.' She also says that she knows people who read the Intergovernmental Panel on Climate Change (IPCC) reports, and the despair they feel, but she chooses not to read them:

> *I don't go there. Maybe I would do more activism if I did read them all, but it seems there's enough suffering going on right now ... that I can turn my attention to.*

She sees that, 'it is not just climate change, but the loss of habitat, and the 'greed of humanity just taking more and more'.

Gerry immediately honed in on the emotional aspects of despair, and the many ways it can impact upon us. As he explained,

> *I've had some fairly intense periods of despair ... I know physically I feel weighed down ... I become non-verbal, close to tears, if not crying. And the worst is when I come to this space of I feel like saying, "Oh fuck it!". I feel like I can't do it anymore, so I'll just turn off. I'll do what I see other people do and just turn off.'*

As he notices, many people cannot deal with the reality of this climate emergency and so shut it out, refusing to see it, or feel it impact.

Lynn is also hit hard, saying,

The despair that I feel has a lot to do with working with racism, and how people don't make that connection to the climate ... it's like an unconsciousness, a holding onto a privilege, that [climate change] is a problem which is over there, for them to deal with ... keeping ourselves separate from the problems facing other people.

Continuing on she explains,

the impact that it has is body-based ... it makes it hard to get out of bed in the morning, I feel tired and lethargic, there's a sense of a weight on me, that I'm repeating myself over and over again, and that I've been doing it for many years... that's my despair. It's a very deep one, and the only way I can get through that is through relationship.'

Lynn identifies here one of the most robust antidotes to despair, which is to connect with others, especially in taking action. In doing so, we are reminded that we are not alone, nor are we powerless.

Community impacts

Others see the impact of despair at the level of community, feeling frustrated that humanity has failed to grasp the significance of the existential crisis facing us. As Garth mused, 'Why don't we just do this better?' Garth's experience, as someone who has had a long-term chronic illness, offers us a fresh perspective on how we might respond to despair. As he goes on to say,

When I got sick, there's a sense where ... I've already gone through my own personal apocalypse, so, it's kinda like, "How much more can climate take from me ... over what I've already lost?", and having already lost that, actually my life is still pretty good. There's all these things that make my life good: having a secure place to live, a loving wife, and having financial security, stuff like that. But also part of the process of having this kind of illness is that I've kind-of given up on planning, I don't have dreams for the future, because ... it's kind of pointless. Well, I wouldn't be able to ... I tried to, but it's frustrating, because I can't actually achieve that. So, in the same way that I don't have dreams for the future, I kind-of don't worry about the future anymore either, because I'm like, well, we'll deal with that when we get to it.

Garth's personal predicament forces him to stay grounded in the present, and able to feel grateful for what he does have, rather than imagining the sorts of doom-filled scenarios often propagated by climate activists. Learning from Garth, we can ask ourselves, 'How might we act meaningfully in the current moment?', trusting that these actions *will* make a difference, even if we cannot see how the future will play out. We will still feel the emotional impacts of emerging disasters, the fires and floods, the destabilisation of governments, the breakdown of societies, but we can choose to engage now, making sure we stay connected, and build relationships with those around us. We can also respond spiritually, using the tools that our spiritual practice offer us, to settle us amid the turmoil, and to notice that despair is always a possibility, but it does not have to consume us.

– silence –

Climate justice

My own despair is felt most keenly when responses to the climate emergency deeply fail the marginalised of the world. Right now (May, 2022) I am noticing and feeling the pain of those who are going hungry, due to rapidly rising food prices, driven in part by Russia's invasion of Ukraine. This is impacting the global poor, and is felt especially hard by First Nations people in remote communities, where the costs of fossil-fuelled freight is a much larger part of the costs of daily necessities.

I am not living on the margins, and do not yet feel many of the impacts of this climate emergency. The truth is that those who have most power to change the trajectory we are on are most blind to the damage we are doing. This is everyone from the world's billionaires, competing in a futile space race, to politicians who ignore the suffering of those whose circumstances of birth made their poverty almost inevitable.

When I was a young adult, I wanted to believe that Australia was not a racist country, choosing to see racism only elsewhere, such as in the legacy of slavery in America. It took older and wiser friends of mine—particularly one of my early childhood education mentors, Dr Glenda MacNaughton—to help me see the truth. She patiently argued with my stubbornness and ignorance, opening my eyes to the realities of racism in our country. It could not have happened without the respect I had for her and thus the realisation that I must be missing something important. This is the reality of white blindness—the blindness that comes from privilege. When you are privileged, you do not even *see* the problems, because those problems do not touch your life. Most people with privilege choose to stay blind, because that is the only way to feel comfortable with their privilege in a world where people starve, are killed, and are humiliated, to preserve the privileges that we hold onto.

Having had my eyes opened I began to see the racism that was all around me, and, once seen, this cannot be unseen. Once begun on this journey, I was forced to realise how much I missed in the world around me, and how much I needed to learn from those who see and name those injustices. One of my first teachers was the poet Audre Lorde—a Black Lesbian who experienced a long struggle with cancer—who saw clearly many of those injustices, around race, around gender, around sexuality and disability. When I first read her work I felt the call of truth within it, but there was much that puzzled me, because of how much I still needed to learn. Rereading her work in the years since then, I am in awe of the ways her wisdom has been decades ahead of the rest of us. I learned that she visited Melbourne, one of the places I have called home, back while I was still in school, in 1985. I was blown away to see her using her speaking opportunity at that time to amplify the struggles of the Wurundjeri people on whose land she was visiting. While acknowledgement of First Nations sovereignty is now more visible, speaking such a truth to a group of white feminists thirty-seven years ago was not a recipe for comfort, but a radical action.

In my work as an early childhood teacher, and later as an early childhood academic, I have embraced what Megan Boler in 1999 called a 'pedagogy of discomfort'. Even working with very young children my job is not always to reassure them that the world is an okay place. Sometimes they need adults to affirm to them that the world can be and often is deeply unfair, and that they can do something about that. When teaching adults studying to become teachers, I definitely wanted them to feel unsettled, and to keep looking for the deeper truths that lie beneath an apparently untroubled surface culture.

In trying to tell a story about climate justice, I know that I cannot rely on what I have seen, but must listen with an open heart to all those who are living with this injustice. Justice has always been skewed in my favour, and climate justice is no exception. I have lived only in countries with high standards of living, and could afford to live in locations less impacted by pollution or erratic weather. To understand this story more fully I have been listening to the words of First Nations people, people of colour, and others who continue to experience marginalisation.

For example, Tyson Yunkaporta, in his book *Sand Talk*, draws on First Nations understanding, as well as the patterns of deep history, to offer all of us a challenging truth. As he notes, city-building cultures, whether in ancient times or today, are never Indigenous cultures, because the nature of cities is perpetual growth. They suck the life out of the lands around them, in ways that First Nations people could not abide. As he explains,

> *Growth is the engine of the city—if the increase stops, the city falls. Because of this, the local resources are used up quickly and the lands around the city die. The biota is stripped, then the topsoil goes, then the water. It is no accident that the ruins of the world's oldest civilisations are mostly in deserts now. It wasn't desert before that. (Yunkaporta 2019)*

Seeing through others' eyes has helped me connect to possibilities for the future that I would otherwise have overlooked. As Habibah explains, one of the most important things we can do is to extend our solidarity, wherever and whenever we can. To recognise the inner light in more of those around us is to feel the increasing urgency of action—we cannot afford to step away from this struggle.

Imagining climate justice

To combat the despair that many of us feel, we need to have a vision of the future we are aiming for, if we are to turn this crisis around. I was made aware of the importance of this by Favianna Rodriguez, who explained how easy it is to catastrophise, to imagine doom-filled scenarios. Yet these terrible visions only paralyse us, preventing us from seeing what we need to do now, what we can do, to change the world around us. She called upon artists and creative people to create visions of the future to inspire and entice us, calling out to our best selves, and invigorating our struggle for a liveable world. Her words called to mind one of my favourite books, *Woman on the Edge of Time*, by Marge Piercy. Writing in the early 1970s, this feminist writer conjured a vision of a future which was communal, egalitarian, connected to the earth, and aiming to live within the carrying capacity of its bioregions. This future world had clearly been damaged by our age of excess, but this was being turned around, and people found space for art, for joy, for love and for satisfying work in the company of others. Having reread this book a number of times since, I realise how much my own imaginings of the future were and are shaped by Piercy's vision, and how much this has sustained me as an activist.

What we fail to notice is that we are being sold—often quite literally—an impoverished and profoundly depressing consumer-oriented lifestyle. Most of this does not satisfy us for long, and almost all of it is made in ways that damage the lands and waters of our beautiful planet. This is no accident. There is no profit in selling people what they really need, because most of this is free. We want to love and be loved. We want times of connection and times of nurturing solitude. We want the work that we do, whether paid or unpaid, to be meaningful and valued by others. We want a world without fear, and a world in which we do not have to be anxious about whether we have shelter or food. These things are possible for all, even with eight billion people on this planet.

I am going to tell you some stories now—these are fictions, built on some of the concrete wisdom of experts on how we could turn this crisis around. Their purpose is to help us have a sense of our destination—where we want to believe we could end up. Collectively, I hope they will help you imagine what a better and more regenerative shared future might look like. I encourage you to imagine yourself listening to interviews with various ordinary Australians living in 2035, if we take up the challenge of bold and regenerative action to address the emergency of a warming planet.

> *My name is Jill. Fifteen years ago, I was on the dole, long term, and let me tell you, it was miserable. Finding a place to rent was a nightmare, and paying bills, when they just seemed to go up and up, kept me awake at night. So these days I feel like I am on easy street. Since the govern-*

19

ment's new climate jobs program, no-one needs to be unemployed if they don't want to be. And it's great. I work with the same crew, and life is always different. For the last few weeks, we have been transforming part of the old stormwater system into an urban wetland and water recycling program. This suburb and surrounding suburbs flooded in the 2022 and 2029 floods, and even though a lot of work has been done already to manage these extreme weather events, it was still touch and go when the floods came here last year. We've been able to make a larger wetlands here because fourteen house blocks were acquired in the flood buyback scheme in 2029, when people decided not to rebuild. We're almost done here, and while the reed beds will take some time to get established, I can imagine what it'll look like, because another crew finished one near my place a few years back, and it is beautiful. So many birds, and frogs, and dragonflies, it breaks your heart. We already know where we're off to next, which is a treeplanting program a couple of hours out of the city. We'll get accommodation for the two weeks we're there, and get to check in on the plants we put in last autumn. All this, and getting a living wage as well! I think my daughter is going to join the program herself, even though she'd be sure to get a regular job anyway, because she loves the outdoor life, and sees how happy I am. I used to be worried sick about climate change, and not knowing what I could do so my kids would have a brighter future. Now I am working to make that change, and I couldn't be happier. Later today we're being visited from one of the local kindergartens who are going to join us in the final plantings before we divert the stormwater in here permanently. It's so great to see the little tackers getting into it.

***

My name is Prija, and I am a teacher at my neighbourhood kindergarten. This is part of the new program that was established a few years back, which guarantees places for children aged two to four, to ensure everyone gets a great start to their education. The best part is they are keeping the kindergartens small, so they feel like a second home for the children. With our fourteen children we have two permanent staff, with a roster of early childhood teaching students doing paid internships, and gaining skills in the progress. We love having these students as they're always full of good ideas, and have heaps of energy for being active with the children. As part of our program we spend a lot of time out in our community, helping the children learn about the sorts of work that people do, and getting an idea of some of the ways our community is pulling together to turn around the climate emergency. This afternoon we're going out with one of our local Elders, Uncle Eddie, to help plant reeds and sedges in a new wetland they're building a couple of streets away. The children have been doing a lot of learning about caring for Country from one of our student teachers, Dale, who is a Wiradjuri and Gomeroi man. He knows so much about the local insects—which the children love. I even hear some of our parents asking *him about bugs they've noticed in their own gardens. When the work crew finish up next week, we'll go back with Uncle Eddie to do a smoking ceremony to open the new wetlands. The children have been writing invitations to local residents, and many of the parents have offered to help put them in people's mailboxes nearby.*

***

Yamandhu marang! My name is Ellen Clements, and I am one of the Wiradjuri Elders on the management committee of our mining project. We see ourselves as world leaders in the field of ecological mining, although we have some friendly rivalry going with a Coast Salish group in British Columbia, who might also make the same claim. We've never been after big profits—we're just happy we have good jobs for our kids to go into. Our aim is to honour our ancestors while also contributing to the resources needed for the booming renewables industry. All our buildings and machinery are electric-powered, drawing mostly on our 500-megawatt community solar project. When I was still working as an engineer, I helped install this. We deliberately put our banks of solar panels on higher poles, to allow room for wildlife to flourish underneath it. Our bilby population is booming, and that means the quolls are too. I don't think most of our older rangers thought they'd be combining their traditional work with solar panel maintenance, but it made sense to us to combine those roles, and the young 'uns totally get it. I had hoped my son Dale would stick around to work on our mining project, but instead he's gone to be a teacher in the big smoke. Perhaps my daughter, Narranderra, will get a job on our new battery plant, when she finishes her chemical engineering degree. Like mother, like daughter, as they say! We're processing our own lithium—lithium we know is being produced in ways that respect the land, and all those who share it with us. Dale comes back every holiday, though, and tells me that the children love his stories about being back on Country. Those children need to know about what our mob are doing, and when he's finished his studies, I'm going to see if he'll come back to work in our Garru early childhood program.

***

Hi, my name's Danyal, and I wanted to tell you about our council's soil remediation program, which I coordinate. I'm a chemistry/biology nerd and never thought I'd end up doing work like this when I was a student back at the University of Iran. When I fell in love with an Aussie and migrated here, I took a postdoc with a team working on algal and fungal bio-remediation, and saw the huge potential in this area. Our team helped develop some new strains which are 30% more effective at dealing with fossil-fuel contaminated soils. That may not sound like much, but trust me, in our field that is revolutionary. I get people trying to headhunt me for work in remote mining sites, but for now Grant's work means we'll stay in the city. I'm told

21

by my colleagues that when councils used to rehabilitate old petrol stations we didn't have many options except to cap it with concrete and turn it into a skate park. The local skater kids loved this, but council officers felt like we were just kicking the problem down the road. At our latest project, which used to be a twenty-four bowser servo, we're just completing our final soil analysis, and have found no traces of heavy metals or volatile organics, after just 18 months of works. That's like the blink of eye in bio-remediation terms. After cracking a bottle of bubbly we're just about to hand over to our council's green waste team, who will dose it up with compost, and hand it over to the local urban landcare group, who tell us they've got tubestock raring to go. We can't wait to see how it looks as a nature-play space for the local primary school.

***

G'day, my name's Grant. When I was growing up in Port Augusta, I never imagined that one day I'd be doing what I'm doing now. I'm part of a social enterprise supporting participatory democracy projects, and I love it. It's such an inspiration being able to show people that their voices make a difference, when some of them have never felt valued or listened to in their lives. I wouldn't be here except for a visit from a group called DemocracyCo back in high school. They ran a day-long program for years 9 to 12, showing us how citizens' assemblies might work to help turn around the climate emergency. Talk about a lightbulb moment! My folks say our electorate was once a Labor stronghold but it doesn't seem like that to me. My whole life it's been a Coalition seat, and so we just get ignored and taken-for-granted around here. Even as a teenager I could see that! I ended up doing work experience for DemocracyCo and have been working in this field on-and-off ever since. I even got to go to Iran to work with a group running direct democracy projects in local neighbourhoods in Tehran. And me just a kid from the sticks, I couldn't really believe it was real! At least I understand deserts, and the challenge of droughts and floods. That's actually how I met my partner Danyal, because he came along to one of the sessions we ran there, dragged along by a friend of his. It was love at first sight for me, but Danyal took a lot more convincing. It's true, I had a fair bit of ingrained racism to unlearn (still do, he tells me!), but after two years of Facetiming and Signal chats, he took the plunge and migrated here. Their loss is our gain. He'd never beat his own drum, but he's a chemistry whiz, and I get so inspired thinking of all the pollution he is helping us clean up across Adelaide. We knew about environmental pollution when I was growing up in Port Augusta, believe me, and I just want to dance for joy at knowing those days will finally be behind us.

***

They used to say "Life begins at 40", but, for me, I reckon "life begins at 70" would have been more accurate. I'm Glenda, and if you'd asked me ten years ago, I would have told you I'd had the most mundane life imaginable. I was part of the Boomer generation and supposedly, we had it all. Yeah right! I was a wage-slave all my life, working at Port Augusta Centre Pharmacy selling people vitamins and incontinence pads for forty-odd years. All I lived for was my garden, and I used to spend all weekend out there tending my roses and my vegie patch. Yes, I own my own house, but only because most of the younger generation left as soon as they could. I lost two husbands to lung cancer, and it used to make me wild when the state government blamed it on smoking, when neither of them smoked a day in their life! We could all see and feel the pollution from the coal power stations in those days, believe you me. I smoked a bit when I was young, and it seemed so unfair that I was the one who outlived them both. The grief almost felled me, I reckon, losing Doug in 2003 and Bob eight years later. By then they were warning us about the air pollution and telling us not to grow vegies around here, but that didn't bring my hubbies back. So when our governments kept supporting coal power through the decades of the climate wars, I felt like I was living a nightmare. When I finally retired, I didn't think I had much to live for. On one of his visits up here, my grandson Grant told me about a regreening project being started by the local Nukunu people, funded through the Treaty Reparations Fund, and profits from the local solar farms. They were looking for volunteers, he said, and I thought, "Well, what have I got to lose? I know how to grow things." Well, how much fun has it been? I have made so many friends, and helped make our town beautiful in the process. We get droughts up here like you wouldn't believe, but Nukunu people have seen it all, and they know which plants will survive. I used to love my rose bushes, but an Eremophila beats them hands down, let me tell you. Those things are bloody hard to kill, just like me, and they flower practically the whole year round, which gladdens your heart. With support from the clever folks at the Arid Lands Botanical Garden, we've learned to propagate many of their plants, and turned the whole place into a garden! We've pretty much regreened most of the obvious spots around town at this point, so now we're starting on green roofs, where the roof will support it. I love it! People think an 80-year-old shouldn't be climbing ladders, but why not? I might as well break a hip doing this, than slipping over in my bathroom. Grant's boyfriend Danyal tells me that Tehran has some amazing gardens, built by the Persians back in the day. We've cooked up a plan to made a Persian garden in my backyard here, using Aussie natives, and I guess I'd better plan to stay alive long enough to make this happen. Who would have thought that dirty old Port Augusta could look like this?

***

23

Greetings. My name is Hao Yu, but you can call me Hao. I am retired now, but for most of my life I had a small business selling solar and other sustainability products, back when these were thought of as 'alternative'. There were times when I almost gave up—years when I would have earned more working as an engineer back in Hong Kong than owning my own business here. When I migrated, they refused to recognise my degree—it's not a lot better for most migrants now. So many hoops to have to jump through it feels like you're starting all over. At university I'd learnt about what was then called global warming, and knew that my children's future depended on something different. We all take this for granted now, of course, but I think the tipping point was back in 2022, when they opened the now-famous Burwood Brickworks shopping - the first serious attempt at sustainable retail big business. It had a massive solar system and rainwater tank, sustainable materials, and basically looked at the complete life-cycle of the building, and how to do it right. Their vision extended beyond just the buildings though, aiming for social change. Mine was one of the businesses offered a good deal to relocate to the "green living" floor of the mall, which brought sustainable retailers of all sorts together in one place. This turbocharged my business, because Burwood Brickworks became justifiably famous, which meant shoppers came from far and wide, many out of curiosity. I went from operating a little shop in Springvale which very few knew about, to being at the centre of a green Mecca! In hindsight we can see that this shopping mall changed the hearts and minds of everyday Australians in ways no-one had predicted. When Chadstone and Westfield started losing customers to the Brickworks, it became a race to the top in the retail sector, and many other businesses started looking closely at how solar electrification could change their own business model. Forget 'free parking'—now everyone expects a charging point at every parking spot! Personally, I still prefer my pushbike. I feel lucky to have been able to pass on my business to my daughters, and I no longer try and keep up with all the amazing products they are now selling. And mostly, I am just relieved that my children and grandchildren will have a future. Not all of it will be as easy as the solar revolution, but at least now we all have a chance.

***

Hello, my name is Ru Shi, and I'm a public transport consultant. I used to work for the MTR system back in Hong Kong, before the Beijing government crackdown on democracy in the early Twenties. I was proud to be part of that organisation, knowing we were the only public transportation system in the world turning a profit, while still expanding our network. I was lucky enough to be sponsored to migrate to Australia by my Uncle Hao. As he said to me then, "You're a smart lady. Australia needs more engineers who really understand railways." My first job was working on the early days of Victoria's Suburban Rail Loop, making sure this was integrated seamlessly with all the existing transit networks, so that when each stage *was finished, it would be off to a flying start. Back then people thought good public transportation only works in high-density cities like Hong Kong, but we've proved that isn't true at all. It's all about the power of networks, and smoothing those interconnections, so the journey is seamless. People aren't stupid. Their transport choices are shaped by convenience and price. If they can get there by train faster than they could sitting in traffic, and cheaper, then of course they will. Plus, they can catch up with the world on their devices while doing so. We saw back in Melbourne that once enough people caught on, the system was embraced by all sorts of people, making it safer and more sociable. And of course, electrified transport is more cost-effective when powered by Australia's world-beating renewables, as well as using less land than roads. Since those early days in Australia I have now consulted with most of the state governments, and am proud to have helped improve the user-experience across all those networks. These days I live in Darwin, because the weather feels like home—though not the crocs! I can do most of my work by telecommuting, and was recently invited to be part of a UN roundtable on 'Regenerative Transport'—if only my poh poh could see me now!*

Friends, we could tell many stories of this near future, and we need those stories as a mud map of what where we are trying to get to. Our Quaker *Advices and Queries* have much to offer us, but for me number thirty-three speaks most clearly to the concept of climate justice:

Are you alert to practices here and throughout the world which discriminate against people on the basis of who or what they are or because of their beliefs? Bear witness to the humanity of all people, including those who break society's conventions or its laws. Try to discern new growing points in social and economic life. Seek to understand the causes of injustice, social unrest and fear. Are you working to bring about a just and compassionate society which allows everyone to develop their capacities and fosters their desire to serve?

The words 'social unrest and fear' speak loudly to our situation, and we do not have to sit idle in response. I have tried to tell you some anecdotes of that just and compassionate society—one that we are called to create together. There are so many stories we could imagine, and bring into being, and I hope you will bring your own creativity to this process.

– silence –

Hope

Hope is something we cannot live without, as illustrated by the famous story of Pandora's box, where all the evils of the world were let loose, but at the bottom, there was also hope. Leaving aside the demonisation of women at the heart of that story—and the sexism it represents—this story tells us of a very human need in the face of calamity. However, as many of the activists I chatted with commented, hope is dangerous if it prevents us from seeing the real state of the world. Hope is sometimes a daydream, something we use to avoid our present and the problems we face. Rebecca Solnit, in her book *Hope in the Dark*, wisely unpacks the idea of hope, reminding us of what it can offer, and what it cannot. As she vividly describes,

> Hope is not a lottery ticket you can sit on the sofa and clutch, feeling lucky. It is an axe you break down doors with in an emergency. Hope should shove you out the door, because it will take everything you have to steer the future away from endless war, from the annihilation of the earth's treasures and the grinding down of the poor and marginal ... To hope is to give yourself to the future—and that commitment to the future is what makes the present inhabitable.

I am here, Friends, as an agent of hope, to shove you out of the door, because these times we find ourselves in demand it. I cannot tell you that we will be alright, that humankind will find our way through this tangled web of international conflict, ecocidal fossil fuel companies, and pervasive inequalities. What I can ask, what I am asking, is that you step out of that door and put your best energies into resolving this crisis. We must do this together, both as Quakers in Australia, but also alongside the countless other inhabitants of our bluegreen planet. This is not a new request. In 2008, the Earthcare statement we adopted as a Yearly Meeting said we would need to 'commit to the demanding, costly implications of radically changed ways of living'. Indeed, I believe many of us have tried to change radically the ways we live, trying to reduce the impacts we make on our local and global ecosystems. Since that time, Australia has endured more than a decade of unnecessary and costly paralysis. Governments of every persuasion have failed to respond effectively to this crisis. We are not alone in this paralysis, nor is it an accident. As a nation-state whose pillaging of First Nations' land has given us access to vast stores of fossil fuels and other valuable mineral wealth, we have been a focus of ongoing manipulation by the 'bad actors' in our community, the wealthy owners of companies who profit from the short- and long-term destruction of our world.

As Elizabeth warns us, 'I'm just not sure that hope will ... lead us anywhere authentic. Perhaps I have a mistrust of hope as a tool of ... jollying us along, or not seeing things as they really are.' We must see the world around us as it is, not how we wish it would be, and open our eyes to the need for action.

What do we have to lose?

In one sense, we have nothing to lose by taking action, except our comfortableness as wealthy-enough people in one of the wealthiest nations on Earth. But in reality we have everything to lose: a stable climate, our food sources, countless species of animals, birds, insects, amphibians, fish, and plant life, a safe political environment, safe drinking water, freedom from war, the lives of our friends and family.

This loss often terrifies us so much that we look away, and distract ourselves with busy lives to avoid having to think of all of these losses. But this climate crisis demands that we remain witnesses to all of the losses, for the sake of those people on our planet whose lives are already being destroyed by this emergency. Close to home, we see those who have been flooded out, whose homes and livelihoods have been destroyed. Many of them were not insured, and indeed many places on our planet are becoming uninsurable – the risks are too great and too frequent to spread the losses. We think further back to the 2019-2020 bushfires, which devastated much of the east coast, including some of the precious remnants of rainforest which had never experienced fire. Many of those properties not only cannot be replaced, but should not be replaced, as we try and learn to live within the constraints of an erratic climate. Some of those ecosystems cannot be replaced, but will need to be restored with plants that may be able to survive inevitable future bushfires.

What do we have to gain?

What we gain is the only habitable planet that we know of in the universe. This planet, our Earth, remains a precious gem of diversity, even now, after decades of extinctions. The more we look closely at our planet, the more diversity we find, because life is abundant and creative, endlessly searching for new connections, new ways to live. We can and must learn to live within the absolute limits of a finite planet and ecosystems that can only adapt slowly to changing climates. We will need to adapt our ways of living, from the wastefulness and thoughtlessness of the fossil fuel era, which has literally burnt up millions of years' worth of planetary savings for little ultimate benefit.

What we have to gain, however, is a more deeply connected, spiritually enriching, and emotionally satisfying existence. In such an existence, we will be learning how to restore ecosystems, make resilient human communities, and provide the necessities of life in ways that are fair for everyone. In such an existence there will be no time for despair, because we will all be needed to put our shoulders to the wheel.

In writing this Backhouse lecture, I have not been able to get out of my mind one of the parables told by Jesus in Matthew's gospel (13; 45-46). It is very short—simply two verses—and goes like this:

> *Again, the kingdom of heaven is like a merchant in search of fine pearls; on finding one pearl of great value, he went and sold all that he had and bought it.*

It would be easy to read this verse and think nothing of it, because it's over so quickly. It has traditionally been interpreted to understand the pearl of great price to be the good news of salvation offered by Jesus' life and death. Yet in this time of climate catastrophe, we can hear another meaning within these verses, which can speak to us both individually and collectively as a Quaker community. The pearl of great price is this precious planet of ours—unique and irreplaceable—a gift of the divine. This planet has supported and continues to support all the lives of creatures and humans we have ever known, every culture and richness of human society, every tiny pocket of beauty and wonder. This parable reminds us that when we discover something of such value, nothing else matters. We can and should give up everything we have, every single material possession, in order to preserve such a treasure. This Earth, our home, is the ultimate treasure, and we have been blind to its inestimable value.

This insight has been thrown into sharp relief in recent years, as some of the world's wealthiest citizens set their sights on Mars. Even if we were to establish human habitats on Mars, such a life would still be infinitely poorer than life in the most damaged landscape of earth. Compared to Mars, even our deserts are teeming with life, with more water, a more breathable atmosphere, and a more hospitable range of temperatures. For a fraction of the cost it will take even to get to Mars, let alone survive there, we could avert this climate catastrophe and raise every single human being on this planet out of poverty. Our earth, this bluegreen planet on the periphery of the Milky Way galaxy, is a world of abundance, even now, with all the damage we have done.

– silence –

Now...and now...and now is the time to act

I could tell you that *now* is the time for us, as Friends, to take action. This will always be true. Just like the old saying— 'The best time to plant a tree was twenty years ago. The second-best time is now'—the best time to act on climate change was thirty years ago, with the *United Nations Framework Convention on Climate Change*, at which time the task would have been so much easier. Yet we have delayed, dithered and debated, while those invested in the fossil fuel system have worked covertly and overtly to undermine even the smallest of changes.

So we come to the current now, in 2022. Some of those whose work I have read talk about the privilege of being alive now, because there has never been a more consequential time to be alive. One hundred years ago, there were less than two billion people on the planet, and our impacts on the global climate were only measurable with difficulty. One hundred years before that, they were almost imperceptible. We live at a pivot point of earth's history when we can impact every lifeform on this planet, from bacteria to blue whales.

I have a request of you all in this awe-filled and awful moment. Can we act collectively as if every life we know is at stake? Can we, Friends?

– silence –

I cannot tell you what exactly this collective action will look like, but I can paint a picture for you based on the collective wisdom of all those I have read and spoken with.

One useful guide is from our Climate Change and Species Extinction Working Group, and the survey they asked us to take late last year. They outlined five main areas in which we can be working for climate action. The first of these is through lobbying people, whether these are our politicians, or the leadership of large companies with significant emissions. The second is through indirect

29

political action, such as letter-writing, or voting, in which we seek to influence through collective effort. The third is direct action, such as being part of protests, rallies, vigils, sit-ins, blockades or whatever, in order to raise awareness, and challenge the status-quo. A fourth is community-building, whether this is in your street, neighbourhood, workplace, or another organisation. This one is often overlooked, but will be vital in our climate resilience, as we've seen in the action taken by flood-ravaged communities in Queensland and NSW earlier this year. Lastly, there is direct habitat restoration, whether this is tree-planting projects, restoration of coastal ecosystems, guerrilla gardening, litter and pollution clean-up, or other direct methods of supporting the health and sustainability of our local environments.

I have not included in this list our individual actions to reduce our impact on the planet, such as driving less, getting solar panels, or reducing our waste. I know that many people have been prepared to make some or many of these changes. However, that individual action can be a trap, focusing on our own lives rather than the greater good. We often take this action because it feels within our power, but it does not do a lot to help the cause of climate justice, whose drivers are far bigger than just the behaviour of individuals. In fact, there is emerging evidence that the concept of a 'carbon footprint' was actually a strategy developed by the fossil fuel industry in the 1990s, to try and shift the blame from their own companies' destructiveness onto individuals. This was a highly successful strategy, undermining our sense of community and government responsibility, and distracting many activists from the real challenges we face as a human community.

I believe that each of us here, no matter what the constraints we face, could imagine ourselves taking action in one or more of those five areas. Your skills, your knowledge, your accumulated wisdom will prove useful in some area of activism—all you need to do is take the first step of connecting with others. As Peri described it to me, in the conversation we had, if we think of these five actions as the lifeboats we need to launch, in the shipwreck of a climate emergency we are facing, then our path ahead is clear. *We do not have to debate whether we want to get on a lifeboat*, we just need to choose quickly which lifeboat we will jump into.

As a Society of Friends, I know there is great goodwill and energy for taking action on this emergency, and certainly we will be stronger and more effective if we act together. The Climate Crisis and Species Extinction working group came into being because our longstanding commitment to Earthcare needs a sharper focus, in these times of global crisis. I cannot dictate what action we take as people of faith, but I do know that we can do more; we *must* do more.

One key challenge I will lay before you, Friends, is a financial one. If I know anything about Quakers it is that we aim to be responsible managers of what money we have, reaching back to the earliest days of Quakerism, when early followers were excluded from many traditional forms of societal support, as a result of breaking away from the powerful church structures of the 1600s. We manage our money carefully, and in the process get gradually more wealthy, without actually having a plan for when and how we will spend this money. There are many worthwhile causes which we support, and have supported in the past. I know this. But we have given only what we feel we can afford to give, saving the rest for an unknown future. So, my challenge, Friends, is that we wake up and notice that there is *no time more critical* than now. Our Society of Friends, along with every other human being on this planet, will never face a challenge this great. We should be spending every dollar we have, **now**, and in the few short years ahead, if it means we can throw our utmost support behind the push towards climate justice.

I urge you to imagine for a moment our Australian Quaker Yearly Meeting as if it were an environmental organisation. Looked at through these eyes, we may not have a huge membership, but we are far better funded that any activist organisation I have ever been a part of. We could offer multiple internships to young climate activists of every colour and background, supporting them financially to extend the impact we can have as Quakers alone. Not only will we be doing climate justice by redistributing money to those who have most need of it, but we can use the depths of activist wisdom I know are present among us, to mentor these activists, increasing the impact they can have. Not every idea they have will succeed. Not every action they support will flourish. But it is their own future they will be labouring towards, as well as our own, and they will see more clearly what is necessary than those of us who have lived too long in the misguided world of extractive capitalism.

Alternatively, we could partner with someone like the Sortition Foundation, to fund a citizen's assembly dedicated to addressing the climate crisis. These are expensive to do well, but are also incredibly persuasive to politicians, because they seek to represent accurately the informed views of every Australian, brown or white, poor or rich, of whatever age, gender or ability. This would draw on Quakers' long-documented work within the United Nations, bringing diverse parties together to talk honestly about our human hopes and fears.

Perhaps we will collectively decide on other action, while taking up this challenge to be bold. Whatever we do, as a Society and as a wider community, we must not delude ourselves that time is on our side. Now, more than any time in our history, we must act, and act powerfully, in all the ways we can possibly imagine. George Fox, the founder of Quakerism, asked us in 1652 'What canst thou say?' Now, with all the words that have been written or spoken about this crisis, I will leave you with a similar question: '*What can we do?*'

– silence –

31

References

Boler, M. 1999, *Feeling Power: Emotions and Education*, Routledge, New York.

Heglar, MA. 2020, We Don't Have To Halt Climate Action to Fight Racism: It's time to stop #AllLivesMattering the climate crisis, viewed 31 May 2022, https://www.huffpost.com/entry/climate-crisis-racism-environmenal-justice_n_5ee072b9c5b6b9cbc7699c3d.

Hopkins, H. 2020, Racism Is Killing the Planet, viewed 12 March 2022, https://www.sierraclub.org/sierra/racism-killing-planet.

Lorde, A. 1988, *A Burst of Light and Other Essays*, Firebrand Books, New York.

Rodriguez, F. 2020, 'Harnessing Cultural Power' in AE Johnson & KK Wilkinson (eds), *All We Can Save: Truth, Courage and Solutions for the Climate Crisis, One World*, New York, pp. 121-128.

Solnit, R. 2016, *Hope in the Dark: Untold Histories*, Wild Possibilities, Third Edition, Haymarket Books, Chicago.

Yunkaporta, T. 2019, *Sand Talk: How Indigenous Thinking Can Save the World*, Text Publishing, Melbourne.

www.ingramcontent.com/pod-product-compliance
Lightning Source LLC
Chambersburg PA
CBHW060751280326

41934CB00010B/2440